REPORT

ON THE

GEOLOGICAL & AGRICULTURAL SURVEY

OF THE

STATE OF MISSISSIPPI.

By EUGENE W. HILGARD, State Geologist.

JACKSON:
MISSISSIPPIAN STEAM POWER PRESS PRINT.
1858.

REPORT.

To His Excellency, William McWillie, Governor of Missisippi:

SIR:—The law makes it the duty of the Geologist of this State, to lay before the Governor, annually, a report concerning the progress of the Geological Survey during the preceding year.

The meaning and intent of this law, has been variously construed, it having been thought by some that the annual report is intended to embrace not merely the *general* statement of the work done and results obtained, but also a description, more or less detailed, of the phenomena observed. The latter plan has been carried out in the report made by the late State Geologist, and is not without precedent in other States. It is obvious, however, that reports of this character can be practically useful only where, as in mineral districts, the detail constitutes in itself, a useful and even necessary guide in rendering productive, resources immediately available. Thus, the partial but detailed reports of the Geological Surveys of Missouri, Kentucky and Tennessee, have been eminently useful in developing the mineral resources of those States.

In view of these facts, it is frequently asked why it is that the reports heretofore published in connexion with the Geological Survey of this State, have done so little towards developing new resources; and that, while enumerating a large number of facts, they are of so little use as practical guides in the utilization of those resources, *few in number*, which have been discovered. In consequence of this failure to fulfill the expectations raised by the brilliant results of the

Geological Surveys of other States, it is frequently urged, and has been promulgated even in the legislative halls, that if this failure is not merely the consequence of incompetency on the part of those charged with the execution of the work, it "is of no use to have a Geological Survey of Mississippi;" and whereas, it had been satisfactorily shown that no minerals were to be found in the State, the survey had better be abandoned at once, to save expense.

It is my object, in submitting to you this brief report, to deal with these objections, and to explain to the people at large, what I have had the honor of discussing with you personally, viz.: the true causes of the defects and shortcomings complained of; and moreover, to suggest the remedies which are plainly indicated, not only by the consideration of the nature of the case, but also by experience in this, and precedents of other States. Having but recently entered upon the discharge of the duties of my office, and having been greatly impeded in my progress in the field, by the want of proper equipments, I should have but few facts entirely new to communicate; and since even these would not be readily appreciated without entering into a lenghthy discussion of previous observations, I shall not for the present dwell upon them. Having, however, been previously connected with the same work for sixteen months, and no progress having been made during the interval, I may perhaps speak with some confidence of its condition and its wants; to which it is the chief object of this communication to call attention.

The Geological features of Mississippi are such as to exclude, according to the accumulative evidence of all other countries, all chances of finding within the limits of the State any natural deposits of metalic minerals, with the exception of iron ore. The latter, although widely diffused over the State, has nowhere been found in quantities sufficient to justify the erection of furnaces; not, at least, so long as they shall have to enter into competition with those of adjoining States, where in some cases inexhaustible beds of ore of first quality occur immediately contiguous to the coal which is required to smelt them. Lead ore has been frequently found, it is true, but always in isolated lumps, obviously transported to the spot by human agency, and no more indicative of a mine underground, than an Indian arrow-head ploughed up in a cornfield, is of a deposit of flint beneath the soil. As for the gold and silver mines discovered broadcast in especially the north-eastern counties of the State, they all resolve themselves either into mines of the ore of copperas, or into intentional frauds. Any one at all familiar with geology and mineralogy, will no more look for metallic mines in the formations of Mississippi, than a whaler would hunt for whales in the river whose name it shares. Experience has in both cases, equally demonstrated the vanity of such expectations.

That "Mississippi is entirely an agricultural State" has been so often said, as to be almost a truism. It would, nevertheless, seem to require further demonstration, with not a few—the same who imagine that all the meaning and object of a Geological Survey is the discovery of mines. Its most important bearing, that *an agriculture and the arts immediately connected with it,* is not very generally appreciated; and still less is the fact, that the benefits which may be expected to accrue to those all-important branches of industry, through the agency of a work of this kind, cannot, for the greater part, be realized until it shall have been carried on, if not to completion, at least to a very advanced stage of progress.

It is not a difficult matter to recognize a metallic ore; to determine its value, and the most successful mode of working a mine, is sometimes the work of a few minutes. But it is not so with soils, the ores from which the agriculturalist extracts his precious materials; or with natural manures, which may serve to sustain the fertility of his soil. While the useful minerals are comparatively few and simple, soils are infinitely varied, and their action on vegetation exceedingly complex. The most experienced eye is unable to judge with certainty of a quality of a soil or marl, or the adaptedness of the one to improve the other, by the eye, or any superficial examination alone. Nothing short of a complete and careful chemical analysis, and extensive comparisons of the results with others, and with previous experience, can give them that practical value and full reliability as guides to the practical man, which in the present state of science he does, and has a right to expect. Investigations of this kind are not a matter of an hour or a day; they require time, extreme care, and the best means of research—not only in the laboratory, but quite as much in the field.

To the minds of those not specially acquainted with the subject, the absolute necessity of extreme accuracy, care and conscientiousness in the execution—especially of the agricultural part of the work—may perhaps be best illustrated by reference to the well known fact, that some of the most important ingredients of soils, the withdrawal of which renders them absolutely sterile, are generally present in them in such minute quantities, that a careless analyst might overlook them altogether. And no less may he utterly fail in detecting the characteristic differences between various kinds of soil, by committing an error which in many other cases where chemistry is applied to practice, would be totally insignificant. An analysis of soil, carelessly made, is useless, and worse than useless. And even a correct analysis may be useless, unless all the natural conditions influencing the soil analyzed, in its place of occurrence, have been correctly observed in the field. And what is true of soils, is equally so in its application to marls, which are intended to improve the soil; if incorrectly chosen, they may do the very contrary.

Again: in a level country, where the underground strata rarely appears on the surface, while their character is extremely variable, it is not an easy matter to ascertain correctly their geological features, which are of the greatest importance with reference to the digging or boring of wells.

Here also, great care and minuteness of observation is necessary.

It is therefore obvious, that in the case of Mississippi at least, a hasty and superficial geological and agricultural survey can do no good. And it is equally obvious, that if it is worth while to make such a survey at all, it must be worth while to have it done *well*.

That it *is* worth while, even the results heretofore obtained may be claimed to have satisfactorily shown. The green sand and marl beds of the north-east and middle portions of the State, the extensive deposits of lignite and of valuable clays, and the numerous mineral springs which have been shown to exist, will in themselves compare favorably with results of geological surveys of other States. Under the natural conditions obtaining in Mississippi, the especial importance of the natural manures just mentioned can scarcely be overrated, when we consider what the discovery of similar deposits has done towards the promotion of agricultural interests in some of the Atlantic States. Our beds of lignite or brown coal, cover a larger area perhaps, than any similar deposit heretofore discovered ; and although not equal in value to good bituminous coal, the time is not far distant when these deposits will prove highly important to many portions of the State. In numerous localities the coal deposit is so near the surface as to require scarcely any other than quarrying operations, in order to obtain a fuel which in some districts of Europe is almost exclusively used for domestic and industrial purposes. And even where the bed is less accessible, its enormous thickness will render the extraction of the material profitable. There is not, perhaps, a State in the Union that can vie with Mississippi in the number and quality of its mineral springs ; and the importance to the people at large, of having these sources of health made generally available by a thorough examination, is manifest. On the other hand, there are regions where the water of ordinary wells is positively injurious to health. The examination of these waters, and determination of possible remedies for the evil, will not be among the least important res... the survey. It will likewise make *generally known* the existence and localities of occurrence of useful materials, as stones for useful or ornamental purposes; of clays, adapted to almost all the various uses of that important material; of limestones not inferior to those of which the imported lime is made; and it will give reliable information as to the quality and best mode of working these and other useful materials. And among the lesser advantages, it will save thousands of industrious individuals the trouble of looking for things which they cannot expect to find, while informing them as to what they may reasonably look for.

But it has been suggested more than once, that, granting the intrinsic value and importance of a geological and agricultural survey, it has been satisfactorily shown that the chief resources it is likely to develop in this State, will be of little use to the *present generation*;—that lands are not yet exhausted, and any one may move to a "fresh place" if he chooses. That as for the lignite, there is plenty of wood to last a lifetime, and more too; and such being the case, that they had rather not go to the expense of having a survey made yet awhile.

It is not likely that a policy so short-sighted, narrow-minded and suicidal, should gain ground in an enlightened community, and in the nineteenth century. As for those who hold and profess such views, their departure for a "fresh place" will scarcely be felt as a loss to the community they desert, and to which they refuse to make themselves permamently useful. Having too common ground whereon to meet such objectors, I shall confine myself to meeting the objections of those who, while willing to do something for the benefit *even* of posterity, still imagine that there is no need of accellerating the survey; and would prefer its indefinite prolongation to the exertion of supplying at once means adequate to insure its speedy completion.

In the first place, it is not a matter of indifference whether have to reclaim an exhausted soil, or simply to maintain fertility by a judicious management of its powers. As between the maintenance and reclamation of soils, the latter frequently requires more than three times the labor and expense, and always involve much loss of time. Some soils, once exhausted, become irreclaimable by any reasonable amount of labor.

Let any one travel through the less fertile districts of our State, and mark the tale told by the numerous deserted homesteads and waste fields, overgrown with blackberries and "broom sedge." He will scarcely escape the conviction that even with us it is not too soon to take measures preventive of an evil which almost laid waste whole districts of Virginia, once the centre of tobacco culture;—which have been and are now being reclaimed by the aid of marls precisely similar to those so abundantly found in our own state. Had those marls been known earlier, Virginia would never have experienced the decline of population and prosperity which at one time created such apprehensions, and resulted in the loss, by emigration, of thousands of energetic and enterprising citizens.

I might also call attention to the fact that the results of the survey will serve as a guide to purchasers of land, since it will inform them, not only of its momentary condition and character, but also of the prospect of permanency of fertility, and the means of improvement. Had the survey been called into existence earlier, it might have saved some money to those unfortunate speculators who, allured by the prairie-like levelness of the Tippah, Pontotoc and Chickasaw "flatwoods," invested their capital in a kind of stock which to their amazement, has remained utterly unproductive.

It would be easy to adduce many more cases where the results of the survey are immediately available; while the ultimate importance is too manifest to be questioned. But, taking it for granted that a sound policy in national economy bids us develop all the resources of a State at the earliest period possible; that a geological and agricultural survey is essential to the development of those resources ; and that when made, it is necessary that it should be *well* done to be of use ; then it remains to be considered whether, taking into account the area of the State, the provision at present existing for the execution of this work, is adequate, and proportioned to the amount of labor to be performed; and whether its failure hitherto to realize the expectations entertained, may not to a great extent be traced to an inadequacy in this respect.

It is my conviction that this failure would to some extent have occurred, under existing circumstances, even if the survey had been in competent and efficient hands all the while. And I may add, that this disappointment is very likely to continue, unless some further provision is made, more adequately commensurate with the magnitude of the work.

The original Act, worded so as to embrace in the geological and agricultural survey, botany and zoology, (or in other words, providing for a complete natural history of the State, almost in the terms of the Act providing for the survey of the State of New York,) made the Principal State Geologist, a Professor of Zoology at the University of Mississippi. He was to spend four months of the year in the field, himself, while his Assistant was to be engaged in the field survey continually, as far as the seasons permitted. And to carry into effect the provisions of this Act, an appropriation of $3,000 per annum, was made.

This arrangement was faulty in principle, inasmuch as during the first years especially, it is indispensible that the Principal should chiefly be in the field himself; and it failed in practice, the Principal being unable to extricate himself from his accumulative University duties. Besides, the subjects to be embraced in the survey according to the original plan, were too numerous by far, to be successfully prosecuted simultaneously with means so limited. Hence we find, in the interesting and ably written volume forming the First Report, by Prof. B. L. C. Wailes, geology occupying a subordinate position, comparatively, among the subjects treated of.

Up to the accession to office of the late State Geologist, in 1854, no analysis had been made in connexion with the survey, (simply because there was no one to make them,) and the data concerning the general geology of the State, were but very fragmentary.

During the years 1854–55, the arrangement just spoken of, continued in force. In January 1856, however, the Board of Trustees passed a resolution relieving the State Geologist of any duties as a teacher in the University, in order that he might devote his whole time to the prosecution of the survey. Up to October 1855, Mr. Harper was without a field assistant, except for a very short period; after that date, and until removed by the Board, in October 1856, I myself held the office of Assisstant; and from that time forward, until March 1857, (when the Act dissolving the connexion thus far existing between the survey and the University went into operation,) the duties of the office of State Geologist devolved upon me by order of the Board of Trustees. From the expiration of my term of office until I was appointed by yourself in March last, the operations of the survey were suspended; since the time during which Mr. Harper held the office after his re-appointment by Gov. McRae, it was occupied by him in the publication of his Report, at New York. By the law detaching the survey from the University, the office of Assistant Geologist was abolished; all the duties thenceforth devolving on the Principal alone.

In reviewing the results heretofore elicited by the survey—prosecuted so far, under circumstances so unfavorable—it cannot but be a matter of regret, that with the exception of my own field notes, all the original records heretofore made, and used by the late State Geologist in the compilation of his Report, (both his own and those of his predecessors,) have disappeared.

It is too well known to require discussion, that however small may be the qualifications of the observer, the records of observations, written down on the spot, is always of value; while in any case whatever, the preservation of this kind of records is of the utmost importance, so long as any investigations connected with the subject are in progress. Nor can any report or theoretical discussion, however complete, replace the original record in this respect.

Concerning the report of the late State Geologist, its merits as a scientific work, have received no higher commendation at the hands of the scientific world, than its position in the scale of usefulness has entitled it to in public opinion at home. It may be proper to mention that although the facts relating to the counties of Tippah, Tishomingo, Pontotoc and Itawamba, together with the diagrams and maps illustrating them, are almost exclusively derived from my field notes of a special survey of those counties, my observations have been perverted and misinterpreted to such an extent, that I am obliged to disclaim entirely any responsibility for the statements given, these having been made to correspond to the preconceived ideas of the writer, rather than to facts. I have to disclaim likewise, the special maps of those counties, which have been enlarged from those accompanying my field notes; not only have some of the lines been arbitrarily changed from those laid down by myself on the spot, but the enlarged scale exhibits a specious pretension to a degree of accuracy not attainable under the circumstances.

In that part of the book in which the materials gathered by myself have been made use of, truth and fiction are so intricately entangled, that it is quite impossible to separate the two by any correction which could be briefly made; and it is a matter of conjecture whether any other part of the book is more reliable than this. Yet this work, which I must consider as entirely unworthy of confidence in all its parts, is all that we now possess to show what observations and results have been heretofore obtained—excepting the matter contained in Prof. Wailes' printed Report, and my own field notes.

We have heard from various sources, that the speedy completion of the survey might be confidently anticipated. In reality, *it is but just begun.* What has thus far been done—a very general geological reconnoisance of the State, and the field-work of the special survey of six or seven counties, would, in the hands of one competent and efficient officer, properly equipped, have been the work of ten or twelve months in the field, when restricting his attention to the objects of ag eological and agricultural survey alone.* Had the arrangement made by the Board of Trustees in 1856 remained in force after the separation of the survey from the University, so that a Principal and an Assistant, each properly equipped, could have jointly prosecuted the work, there would have been a chance of reasonable progress; more especially when (as has been the case for some time,) it was understood that, for the present at least, geology and agriculture were to be the sole objects of the survey, to the exclusion of any more special pursuit of the less important, though not less interesting branches of botany and zoology.

But it is no more reasonable to charge a single person, no matter how competent and efficient, with the execution of a work like this, than it would be to employ a single workman to build a house. Every one knows what disadvantages the latter would be laboring under; he might work a lifetime, having to perform himself all the particulars of making and laying brick, sawing lumber, splitting shingles, etc., and after all, the edifice will neither be as perfect as it might have been made with very little assistance from others, nor will it have been of any use until completed; and what is more, every one, and most of all the workman himself, would be out of patience with it. And no one would suppose that the capital so employed had been profitably invested.

Now, the case of a geological and agricultural survey of a State like Mississippi, is precisely analagous to that of the house just cited. The labors required in its performance

*It must be borne in mind, that it is only during eight months of the year, *at farthest*, that field-work is practicable even in the most favored parts of Mississippi.

are scarcely less various in their character ; the simultaneous co-operation of several persons is of even greater importance ; and the disproportion between the means employed, and the work to be done, is quite as glaring in the eyes of those acquainted with the nature of such surveys. But in order to make this apparent to all, it may be well to cite the precedents of some of the other States in which geological surveys have been ordered.

In most cases, certain sums were provisionally appropriated for the purpose of those surveys ; while the specialties of their execution, and among these the determination of the force to be employed, were left, more or less, in the discretion of the Governors and principal geologists. The work could thus be placed under the conditions most favorable in the judgment of *competent* persons, to its advancement, under existing circumstances ; they were thus prosecuted as far as the means permitted, additional appropriations being thereafter granted as they became necessary.

Thus, the State of New York, in. 1836, appropriated the sum of $104,000 to carry into effect the provisions of an act of which, be it remembered, the act providing for a geological survey of Mississippi is almost a literal copy. The State *exceeded by Mississippi in area*, was subdivided into three districts, in each of which not a single person, but a full corps, and sometimes several of these, were engaged in the operations of the survey. Even, thus the field work extended over a considerable number of years, nor has the publication of the results been quite completed even at the present time. Subsequent appropriations have swelled the amount expended (so far as I have been able to ascertain from the incomplete documents at my disposal,) to more than $200,000 —it has been currently reported to exceed a quarter of a million, at the present moment. Well may New York be proud of that page of her history which bears the record of her survey ; the results of which—as laid down in that magnificent work, the Natural History of New York—aside from their great practical importance, have proved to the world more convincingly than thousands of Annual Commencement and Independence Day orations, that republican institutions are not *necessarily* unfavorable to the progress of science ;—*that science* which though still affected to be held in contempt by not a few who call themselves "practical men," has but so recently achieved one of the proudest triumphs of the age—the Atlantic Telegraph!

As yet, the example of New York stands unrivalled on either side of the Atlantic. Six years ago, Mississippi gave promise of following in the footsteps of her senior sister; but she failed to redeem that promise in practice. Taking as a basis the above estimate of $200,000, and considering that like effects invariably require like forces to produce them, the annual appropriation of $3,000 made by Mississippi, would require to run for 66, *say sixty-six years*, in order to complete the survey in accordance with the act creating it!

But although the letter of the law of 1852 still remains unchanged, it has been practically assumed for some time past, (as before observed,) that the more immediate objects of a geological and agricultural survey are the first to be attained. Let us compare notes, then, with some of the States in which surveys of this kind have been ordered. Alabama, Kentucky and Missouri have adopted the New York plan of appropriating from time to time such sums as might be found requisite for the successful prosecution of the work. Missouri began, a few years since, by making an appropriation of $20,000 ; since then has had no less than three parties in the field at any time, besides a chemist constantly employed in the laboratory work. Being in skillful and efficient hands, the results of the work have, under this system, been such as to insure the requisite appropriations, whenever needed. The same has been the case in Kentucky; her survey began in 1854, and now almost completed, was carried on by three corps simultaneously, appropriations sufficient to sustain them being made as required. The Alabama survey, at first advancing slowly, with limited means supplied by the State University, was so far advanced by an appropriation of $15,000, placed in the hands of a Tuousey, that at the period of the untimely demise of that eminent man, it might be considered as half completed, although in his labors he had been aided by but one Assistant. Unfortunately, however, in this case, the appropriation failed to be renewed in proper time, and during the consequent suspension of operations (not even the printing of a report having been provided for,) a large part of the results so far obtained, were buried with the observer. It must be recollected, that when a work of this kind is suddenly stopped at a point, when, *if prosecuted*, half of the labor might be considered as having been performed, that first half will not by any means contain a proportional amount of useful information; no more than if we arrest the manufacture of cloth at the point when the yarn is ready for the loom, we can use *that* material as an inferior kind of cloth!

It may be that the evils attendant upon the uncertainty of the renewal of an appropriation *necessarily* subject every few years, to discussion in the political arena, are among the reasons which induced other States to adopt a different form of appropriation; which, by rendering a repeated special discussion unnecessary, should make the continuance of a work of acknowledged importance to all, less liable to obstruction by every variation of the political wind-vane. But in most cases, while avoiding Scylla, they have fallen into Charybdis. They intended to shield the young tree from being prematurely cut off; but while successfully preventing this, they forgot to supply to it such nourishment as should enable it to grow *at all*. Both in Tennessee and Mississippi, where this policy has been pursued, the annual appropriations are insufficient to secure the completion of their geological surveys within a decennium or two. Tennessee has done even less than Mississippi. Yet we see (and this fact has added to the dissatisfaction of the Mississippians,) that the geological survey of the former State stands high in the good graces of the public. With a due appreciation of the high qualifications of the learned gentlemen now in charge of that survey, I must observe, that the natural conditions existing in Tennessee are as decidedly *favorable* to the appreciation of the geological survey by the public generally, as in Mississippi they are *unfavorable*, for reasons before stated. Had we been able to proclaim the discovery in a Ducktown mine, of quarries of exquisite marble, or inexhaustible beds of stone coal, our short-comings might have been covered with the broad mantle of charity. But the dingy exterior of our marls, lignites and limestones, is little calculated to dazzle the public eye.

Our neighbor State of Arkansas, has done somewhat better, by entering the field with an annual appropriation of $4,000. This sum, having been appropriated *from the outset*, and under circumstances peculiarly favorable, appears to be adequate to insure to that survey a fair rate of progress, in the hands of its distinguished Principal, aided by one Assistant.

It may be observed that with the single exception of Tennessee, no other State has made as small a provision for its geological survey as Mississippi; notwithstanding that, as before set forth, her survey requires more especially, great and time-costing accuracy throughout, and will not be satisfactory to any one unless speedily completed.

It would perhaps be desirable, and a saving both of time and money, were Mississippi to adopt provisions similar to those made by Missouri and other States. Had the sum heretofore expended on the geological survey of this State, been placed at once in the hands of a competent man, with powers (under strict accountability,) to make such disposition of the funds as, in the Governor's and his discretion, should seem best adapted to further the survey, the latter might, without further appropriation, have been more than half completed long ago.

But I do not mean to propose, at the present time, innovations which might prove distasteful to a good many,—especially those who prefer spending in small sums twice the amount which, if at once appropriated, would finish the work. The public does not appear to be prepared to receive favorably such a proposition, and under existing circumstances, it is perhaps best to adhere to the old plan, with such modifications only as the interests of the State imperiously demands, and without which, no real and satisfactory progress is attainable. *The minimum is the provision for an Assistant, and for the establishment of a suitable laboratory.*

In regard to the latter, it may perhaps be surprising to those acquainted with the Legislative Acts concerning the survey, that the $1,200 appropriated by the Legislature, in 1857, for laboratory purposes, should not have been sufficient to accomplish that end—as they undoubtedly would have been, had they been judiciously managed. But the vouchers filed in the Auditor's office, by the late State Geologist, as well as the stock on hand (now deposited in one of the front rooms of the Penitentiary,) show that scarcely more than one-half of the sum appropriated has been applied to the purchase of articles really useful or necessary for an analytical laboratory. The rest, (except $54 11 still remaining in the Treasury,) has been expended in the purchase of promiscuous articles adapted to exhibition, amusement, and other purposes foreign to the survey; (among the larger items, I may mention $212 50 for *two* microscopes, and $73 75 for *meteoroligical instruments.*) And from among the articles mentioned in the vouchers, and certified as having been received by the late Geologist, a number, to the value of $110, are nowhere to be found. Many articles of first necessity, have on the other hand, been altogether omitted; and the appropriation has been exhausted to within a fraction, without there being the least provision for the fitting up of a room with bare walls, to make it answer the purposes of a laboratory.

$500 more will, at the lowest estimate, be required to put the survey laboratory in working order. A part of this sum might, it is true, be derived from the sale of such of the superfluous articles as may be saleable. But these can at best be sold only at a sacrifice, and not very readily at that. The proceeds might, if necessary, at any time be charged to the current survey appropriation.

The salary of the Assistant, heretofore, had been fixed at $1,000 per annum. Experience proves that this sum is insufficient to form a permanent inducement to any competent man; and a continual change of persons has been the consequence of this inadequacy. Now, all changes of this kind are not only vexatious and involve a serious loss of time, but each one is in itself a positive pecuniary loss to the State; for the simple reason that full efficiency on the part of the incumbent, whether Principal or Assistant, requires a certain amount of personal and local experience, which can only be obtained by connexion with the work itself, for some length of time. I suggest, therefore, that the salary of the Assistant be fixed at $1,500—the same as in other States.

In the present stage of the survey, the duties of the Assistant would not, as a general thing, be in the field, but in the laboratory; hence, no additional field outfit would, for the present, be required. The appropriation now existing, of $1,000 per annum, for current expenses, is not however, sufficient to defray the expenses of both laboratory and field work, when carried on simultaneously; an additional appropriation of $300 per annum, will be nenessary for this purpose.

The permanent annual appropriation required for the survey, under this arrangement, exclusive of any contingencies, would therefore amount to $4,800; besides which, $500 are at present required to complete the chemical laboratory.

The act of the Legislature, disconnecting the survey from th e University, provides that "the State Geologist shall keep his office in the city of Jackson;" and furthermore "that the State Geologist may occupy as a laboratory the two front rooms in the second story of the Penitentiary, and shall be allowed the assistance of one convict, to be named by the inspectors, to aid him in keeping his apparatus in good order."

There scarcely appears to exist, in the nature of the work, any reason why the locality where the in-door work of the survey is to be carried on, should be fixed by law. The headquarters of the geological survey of Arkansas and Kentucky, for instance, are at the residence of the Geologist to those States, Dr. D. D. Owen, in Indiana. It is true that, all other things being equal, it would be natural that a department under control of the Executive, should be located at the seat of Government. The fact that in most cases the collections resulting from geological surveys have been ordered to be ultimately deposited at the State capital, would alone furnish a sufficient reason, since the removal of collections of this kind can rarely be effected without some injury to delicate specimens, and always gives a good deal of trouble.

In the present instance, however, it is ordered that the specimens collected be divided between the State University and a State collection to be formed at Jackson. It becomes, therefore, from the outset, a matter of indifference in which direction one-half of the collection may finally require to be moved. Almost all the specimens heretofore collected have been deposited at the University; and were the removal of headquarters necessary at the present stage of the survey, it would be quite impracticable to effect the subdivision of the collections, save to a very small extent. For the work of examination and determination remains to be done yet, and it is of the utmost importance in this study, to have the materials as full and complete as possible. Practically, it would become necessary to remove almost the whole of the collections, involving a great expenditure of time and labor, and the destruction of many specimens—the majority of these being of a character unusually delicate. But worse than this, it would deprive the University for several years to come, of the use of a collection illustrating the geology of the State; or in other words the opportunity of ocularly demonstrating to its sons the very phenomena which they will have to deal with in practice.

For the elucidation and elaboration of the results which still lie dormant, as it were, in these specimens, (comprising nearly one-half of the whole work,) it is indispensable that there should be at command, if not a library, at least a considerable number of books not usually found in general libraries. The greater part of the books so required, the expense of which is not inconsiderable, were purchased by the University during its connexion with the survey, with a special view to the necessities of the latter; whilst they cannot be found in the State library at Jackson, nor so far as I know, in any other library in the State. Were the head-quarters of the survey to be removed from Oxford, it would become necessary to purchase these books once more, and a special appropriation would be required for the purpose.

In view of these objections, nothing short of very considerable advantages to be realized by the removal, could justify the same. But there is not actually a single one to be gained, unless indeed, that the State of Mississippi could not afford to provide for its geological survey, other accommodations than two rooms in the Penitentiary, or other assistance than that of a convict.

The rent of rooms suitable for the purposes of the survey would, in any case, be but a trifling expense; but even this extra expenditure may perhaps be avoided in the present, the board of Trustees having, at their last meeting, expressed their willingness that such rooms in the University buildings as, in the discretion of the President, were not required for University purposes, might be appropriated to the use of the survey. There appears to be little doubt that the collections can and will be thus accomodated, it being in the interest of the University to have them near at hand. An apartment for a laboratory may not be as readily found, there being special requirements to render a room suitable for that purpose; but the difficulties to be overcome in this respect, at Oxford, will scarcely be greater than those existing on the second floor of the Penitentiary.

Should you think fit to recommend, and the Legislature to authorize the changes here proposed, I might hope to be able to lay before you, next year, a summary of results, which should satisfy the sceptical as to the importance of the survey, and the advantages offered by an energetic prosecution over the dilatory system thus far prevailing. Should no such change, however, be effected, I shall still continue to do my duty, although with a consciousness that the severest toil will be unavailing in furthering the work either to my own satisfaction, or that of the public; and that the people of Mississippi will be late in reaping the benefits which the prompt and thorough execution of their geological surveys, has conferred, and is conferring upon other States.

With the highest respect, your obedient servant,

EUGENE W. HILGARD, *State Geologist.*

JACKSON, August 1858.

www.ingramcontent.com/pod-product-compliance
Lightning Source LLC
LaVergne TN
LVHW020634110826
845149LV00004B/1182

* 9 7 8 1 4 1 8 1 9 1 5 8 0 *